# Curing and Fermentation of Cigar Leaf Tobacco

by US Dept. of Agriculture

**with an introduction by Roger Chambers**

This work contains material that was originally published in 1899.

This publication was created and published for the public benefit, utilizing public funding and is within the Public Domain.

This edition is reprinted for educational purposes and in accordance with all applicable Federal Laws.

Introduction Copyright 2018 by Roger Chambers

# *COVER CREDITS*

**Front Cover**
*Nicotiana tabacum 0842* by Dominik Haenni
[CC BY-SA 3.0 : https://creativecommons.org/licenses/by-sa/3.0/deed.en],
via Wikimedia Commons

**Back Cover**
*From interior.*

**Research / Resources**
*Wikimedia Commons*
www.Commons.Wikimedia.org

Many thanks to all the incredible photographers, artists,
researchers, biographers, historians, and archivists who share
their great work via the Wikipedia family.

**PLEASE NOTE :**
As with all reprinted books of this age that are intended to perfectly reproduce the original edition, considerable pains and effort had to be undertaken to correct fading and sometimes outright damage to existing proofs of this title. At times, this task can be quite monumental, requiring an almost total rebuilding of some pages from digital proofs of multiple copies. Despite this, imperfections still sometimes exist in the final proof and may detract slightly from the visual appearance of the text.

**DISCLAIMER :**
Due to the age of this book, some methods or practices may have been deemed unsafe or unacceptable in the interim years. In utilizing the information herein, you do so at your own risk. We republish antiquarian books without judgment or revisionism, solely for their historical and cultural importance, and for educational purposes.

# Self Reliance Books

Get more historic titles on animal and stock breeding, gardening and old fashioned skills by visiting us at:

# http://selfreliancebooks.blogspot.com/

# *DISCLAIMER*

This book was written in an age when little was known about the ill effects of tobacco.

The material presented herein is intended to be strictly for educational purposes with the purpose of enlightening readers about the historical uses of tobacco. Publication of the material is neither an endorsement, nor a criticism of its contents. This book is presented as part of large series of educational material on the history and cultivation of tobacco.

As the reader, please consider it your duty to consult with a medical doctor before utilizing tobacco. It is also the reader's duty to become familiar with local, state, provincial and federal laws relating to the growing of
tobacco.

As the author, publisher and retailer cannot control how the reader utilizes the historical information presented in the pages herein, they hereby disclaim any liability to any party for any loss, damage, disruption, death or other liability that may be incurred by the reader's misuse of this material.

# introduction

Here at **Self-Reliance Books** we are dedicated to bringing you the best in *dusty-old-book-knowledge* to help you in your quest for self-sufficiency and independence.

We're so pleased to bring you this old title on curing and fermenting tobacco. These old reports and bulletins put out by the USDA are very popular. It should be said, though, that some of the information is best looked at in the historical aspect, due to the obsolescence of some practices or methods.

This special edition of **Curing and Fermenting of Cigar Leaf Tobacco** was written by the *U.S. Department of Agriculture*, and first published in 1899, making it well over a century old. It is also known as **Report No. 59**.

This super-short, fast read features sections on *The Curing, The Sweating, or Fermentation Process, The Cold Sweat, Aging, or After-Fermentation, The Petuning of the Tobacco*, and more.

Another great USDA publication that is a must-have for the libraries of all those interested in the historical aspect of the Tobacco Industry.

~ *Roger Chambers*
*State of Jefferson, March 2018*

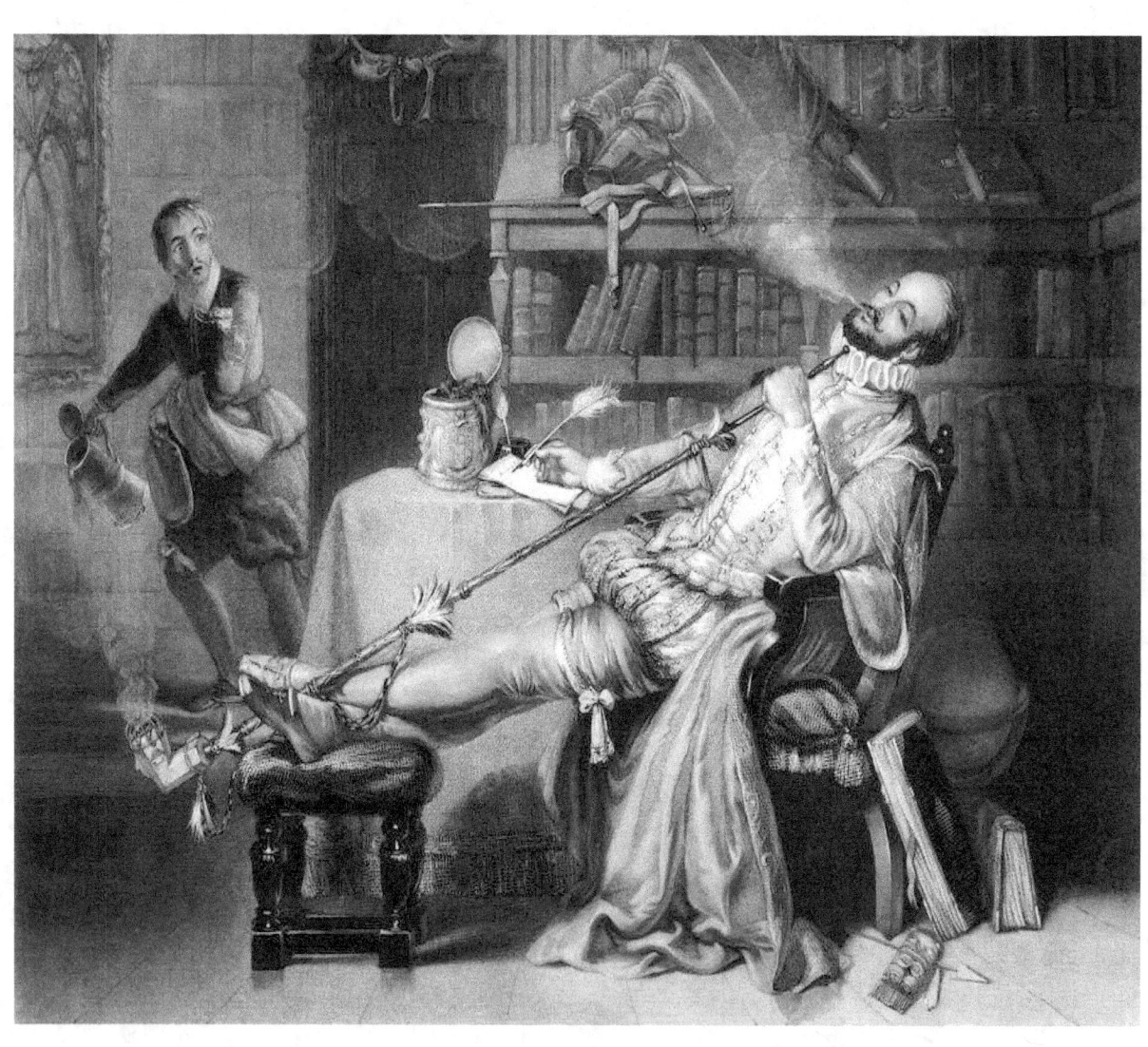

# LETTER OF TRANSMITTAL.

U. S. DEPARTMENT OF AGRICULTURE,
DIVISION OF SOILS,
*Washington, D. C., February 19, 1899.*

SIR: In accordance with the recommendation in my annual report for 1898, approved by you, and the authorization of Congress, a comprehensive line of tobacco investigations, to extend and supplement the tobacco soil investigations of the Division of Soils, has been undertaken. The work includes the mapping of soil areas, studies in fermentation, improvements in breeding and selection, investigations of the conditions of growth and manipulation in foreign countries, and the question of supplying tobacco to foreign markets.

In examining and classifying the soils of the principal tobacco districts of the United States certain facts developed in regard to the commercial value of the crop from certain soils which could not be clearly understood or explained without a further investigation of the methods of curing, fermenting, and handling of the tobacco, and possibly also of breeding new varieties. Only in this way could the full value of the soil work of this Division be shown. As soon as it was definitely determined that the work could be undertaken, I requested the Chief of the Division of Vegetable Physiology and Pathology to detail an expert to investigate the curing and fermentation of tobacco, this work naturally pertaining to his Division. In accordance with this request Dr. Oscar Loew was detailed to carry on the investigations, and at once went to Quincy, Fla., where he spent some time during the fermentation season.

Other Divisions have also been asked to cooperate in a similar manner in other phases of the comprehensive investigation. In view of this extensive cooperation it is proposed to issue a series of reports on tobacco investigations, to which all the Divisions of the Department may contribute matter pertaining to the subject.

Dr. Loew's discovery of the real cause of the fermentation of cigar tobacco, as remarked by Mr. Galloway in submitting this report, can not fail to prove of great scientific interest and economic value, and will unquestionably modify the methods of curing and fermenting when the investigation has been carried further and the conditions and principles of the process are better understood.

This treatise, which is more or less technical, will be followed by a more popular one giving the substance of Dr. Loew's investigations in connection with some temperature studies which have been made in the fermenting piles of tobacco in Florida and Connecticut. I respectfully recommend that the manuscript herewith submitted be published as Report No. 59 of the Department.

Respectfully,

MILTON WHITNEY,
*Chief of Division,*
*In Charge of Tobacco Investigations.*

Hon. JAMES WILSON,
*Secretary of Agriculture.*

# LETTER OF SUBMITTAL.

U. S. DEPARTMENT OF AGRICULTURE,
DIVISION OF VEGETABLE PHYSIOLOGY AND PATHOLOGY,
*Washington, D. C., February 19, 1899.*

SIR: I respectfully submit herewith the manuscript of a bulletin prepared by Dr. Oscar Loew, of this Division, on The Curing and Fermentation of Cigar Leaf Tobacco. The work on tobacco has been carried on in accordance with the plan of cooperation recommended in your report to the honorable Secretary of Agriculture for 1898. The investigations have involved bacteriological, chemical, and chemico-physiological studies, and the interesting results obtained will, it is believed, open the way for further work along important lines.

The chemical work has been carried on in the laboratory of the Division of Chemistry, and we are greatly indebted to Dr. H. W. Wiley, the Chief of that Division, for facilities furnished. At Quincy, Fla., Dr. Loew received much information and valuable assistance from Mr. Henry Storm and Mr. W. M. Corry, second vice-president and general manager, respectively, of the Owl Commercial Company, which has a very large tobacco plantation at that place and which has done more than any other agency in developing the tobacco industry in Florida. We wish to express our thanks to these gentlemen for their kindness.

Respectfully,

B. T. GALLOWAY,
*Chief of Division.*

Prof. MILTON WHITNEY,
*Chief, Division of Soils,*
*In Charge of Tobacco Investigations.*

# CONTENTS.

| | Page. |
|---|---|
| Introduction | 9 |
| The curing | 10 |
|     Decrease of protein | 10 |
|     Regulation of heat and moisture | 11 |
|     Flavor | 12 |
|     Color | 12 |
|     Ammonia | 13 |
| The sweating or fermentation process | 13 |
|     Rise of temperature | 14 |
|     Oxidation | 14 |
|     Losses | 15 |
|     Development of gases | 16 |
|     Starch | 16 |
|     Sugar | 16 |
|     Tannin | 17 |
|     Fiber | 17 |
|     Ashes | 17 |
|     Nitrate | 17 |
| The cold sweat, aging, or after-fermentation | 18 |
| The petuning of the tobacco | 19 |
| The bacterial fermentation theory of Suchsland | 20 |
| The oxidizing agency in the fermenting tobacco leaf | 23 |
|     Views on the physiological functions of the oxidizing enzyms | 26 |
|     The tobacco oxidase and peroxidase | 27 |
| Summary | 33 |
| Recent foreign literature | 34 |

# CURING AND FERMENTATION OF CIGAR LEAF TOBACCO.

### INTRODUCTION.

The production of tobacco adapted to the different market demands has become a prominent factor in national economy. Of particular importance is the production of superior cigar leaf tobacco. The filler leaf of a cigar must above all things have a good flavor, good aroma, and good burn. In the wrapper leaf, however, still other qualities come in, such as elasticity, pliability, size, shape, color, size of the veins, the fineness and peculiar grain of the Havana type, and the smooth silkiness of the Sumatra.

Little is known of the chemical properties of the leaf, especially of those which contribute to the flavor and aroma. It is probable that the actual amount of nicotine is relatively unimportant, and it is certain that the excellence of the leaf and its adaptation to market demands is not dependent, except in a very general way, upon the amount of nicotine. It has long been known that certain of the potassium salts, especially potassium chlorid, can not be used at all for the production of high types of cigar tobacco, as they give the leaf a poor burn. It is furthermore an old experience of tobacco growers that excessive nitrogenous manuring tends to produce a large leaf, of inferior quality, containing an increased amount of nicotine. If the prime object of tobacco culture were the production of nicotine, as the prime object in raising sugar beets is the production of sugar, then the rational procedure would be to furnish an excess of nitrogenous manures, but nicotine alone does not make a good cigar tobacco any more than alcohol alone would make a good wine. The substances producing the flavor and aroma, therefore, although probably present in minute quantities, are much more important than the actual percentage of nicotine found in the cured leaf.

Whitney[1] has shown that tobacco suited to our domestic cigars is grown only upon certain soils and under certain climatic conditions. It appears, therefore, that the leaf capable of being converted into a cigar leaf through the ordinary processes of curing and fermentation must

---
[1] Bull. No. 11, Division of Soils, U. S. Department of Agriculture.

possess certain characters. A fresh leaf has no specific taste, nor has it any specific odor, but the finished leaf has a sharp, saline taste and a characteristic odor.

From the time the tobacco leaf is gathered in the field until the manufacture of the cigar and even afterwards a series of highly interesting changes take place in the leaf, as a result of which the characters of the finished leaf are developed and fixed. There are three stages in these changes, viz, (1) the curing process; (2) the sweating, or fermentation; and (3) the cold sweat, after-fermentation, or aging, as it is variously called.

## THE CURING.

There are two periods in the curing process: The first period, in which the cells of the leaves are still alive and induce processes of metabolism; and the second period, in which the cells have died and the chemical changes have therefore no connection with the living protoplasm. In the former period, which may last only a few days (longer with the ribs), the starch content is dissolved and the sugar formed is partly consumed by an increased respiration[1] and partly transported to the ribs, where, as Müller-Thurgau has shown, starch may be formed again. In the latter period the enzyms alone are active.

*Decrease of protein.*—With the consumption of a large amount of the sugar a state of inanition or starvation sets in, and the reserve protein is attacked by an enzym, trypsin-like in character, the action of which will continue after the death of the cells. A cold-prepared aqueous extract of a fresh leaf will show albumin on the addition of nitric acid and warming, while the cured leaf does not give this reaction. The reserve protein and a certain albuminous portion of the nucleo-proteids of the protoplasm will thus finally be split and transformed into amido compounds and bases, only the remaining nucleins resisting, hence the decrease of protein matter in the curing and fermentation process will stop at a certain point. Such proteolytic processes proceed not only in plants exposed to darkness, which means their starvation or inanition, but also in all cases where reserve protein must be dissolved to enable further development, as in germination or development from bulbs.

It is in full accordance with physiological principles that when cells are in want of nourishment they produce a larger amount of enzyms than when well nourished. This explains why tobacco leaves killed immediately after being gathered will show imperfections when after having been moistened they are subjected to the curing process. The enzyms that have been produced during the inanition state of the

---

[1] The respiration of a pile of such fresh leaves may soon lead to a considerable and even injurious rise of temperature, as in the respiration of germinating barley on the malting floor. A moderate rise is often intentionally brought on, as it hastens the curing. Sometimes this rise of temperature is called sweating, although the cause here is a different one from the true sweating, or fermentation, following after curing.

cells, however, will naturally remain active after the death of the protoplasm from starvation has set in.[1]

Considerable variation has been found in the total nitrogen content of the fresh leaves, as well as in the amido nitrogen content of the cured leaves. The amount of the former may vary in American, Greek, and German tobaccos from 2, 3, or 4 per cent to 8 per cent of the dry leaf. One-third of this and even more can turn into amido compounds in the curing process.

*Regulation of heat and moisture.*—Further changes, relating to color and flavor, set in with the death of the cells. However, it requires a most judicious regulation of the moisture, temperature, and ventilation of the barn where the tobacco leaves are hung up to obtain those changes which characterize cured tobacco of a superior quality. This curing process may last four weeks or even much longer.[2] When the weather is too dry all the chemical changes in the leaves come to a premature stop, but on the other hand when it is too moist the danger of mold development arises. In the former case the barns must be eventually kept closed and water introduced, while in the latter case careful application of heat may be resorted to.

An interesting experiment in curing by artificial heat has been described by E. H. Jenkins.[3] Some farmers have tried the burning of sulphur with the intention of killing the mold spores by sulphurous acid, but this requires the utmost precaution, as the leaves themselves might easily be injured and even all further action in them stopped.

Sometimes mold fungi will develop unnoticed in the stems, appearing distinctly later on, when the sweating operation has begun. All diseased leaves must be discarded before fermentation begins in order to avoid further damage by the spreading of the fungi. Tobacco growers in Florida recognize the white mold, the yellow mold, the blue mold, and the stem-rot mold, the latter being the worst and causing much damage. Sturgis has described a bacterium causing pole burn of tobacco,[4] and further determined the fungus causing the stem rot to be *Botrytis longibranchiata*.[5] Jenkins reports that the pole burn disease "may destroy a portion or even the whole of the harvested crop within forty-eight hours after the time when the trouble is first noticed."

---

[1] It is somewhat difficult to prove the presence of diastase in healthy normal leaves, as very small quantities may resist extraction.

[2] The drying, or curing, for good cigar tobacco requires about as much time in America as it does in Europe. Tscherwatscheff, a Russian, has described the American method as requiring but four days with applications of artificial heat (Landw. Jahrb., 1875). What he had seen, however, was nothing but the preparation of light-colored cigarette tobacco as practiced in North Carolina, Virginia, and Kentucky. In curing cigar tobacco fire is resorted to only when damp, foggy weather prevails for a long time.

[3] Conn. Agr. Expt. Sta. Ann. Rept., 1897; *Ibid.*, 1892, p. 38.

[4] Conn. Agr. Expt. Sta. Ann. Rept., 1891.

[5] *Ibid.*, Sturgis's list of tobacco diseases.

*Flavor.*—The development of the flavor of cured tobacco has not yet been explained. At first a decided flavor of cucumbers[1] is generated, which later on is entirely replaced by the rank and common straw smell of cured tobacco, giving rise finally to the superior tobacco flavor developed by the sweating or fermentation process.

*Color.*—As regards the brown color of cured and fermented tobacco, there can hardly be any doubt that not only one but several compounds contribute by their chemical changes to its development. Of course the first supposition would be that the tannin, by being changed into a phlobaphene (a brown product), is the principal cause.[2] Thus, for example, in the autumn, when the leaves of oaks and of various other trees containing tannin die off, a brown coloration sets in. But the intensity of the brown color of the fermented tobacco leaf does not run parallel to the different concentration of the tannin in the cell systems of the leaf. A healthy tobacco leaf was placed with its base in a dilute solution of ferrous sulphate (about 1 per cent) for from twelve to fifteen hours, at the end of which time this reagent had risen to the tip of the leaf, thereby partly killing it. A reaction in the form of a black color appeared, principally in the epidermis and to some extent also in the mesophyll, but not at all in the vascular bundles. This black coloration seemed to be restricted to the chloroplasts.

The epidermis of cured leaves, however, contains the least amount of coloring matter and is sometimes entirely devoid of it, with the exception of the gland hairs, while the mesophyll cells always contain a brown substance in irregular-shaped or rounded masses. The principal part of the brown matter, however, is in the veins of the leaves and even the most minute ramifications of the vascular bundles appear to be a much darker brown than the neighboring mesophyll cells. The circumstance that the veins contain less nicotine than the rest of the leaf also militates against the view that the coloring matter is principally due to the oxidation of nicotine. However, there occurs in the veins a bitter principle that does not seem to occur in the rest of the leaf, and perhaps this may contribute to the color.

It is easy to show that several compounds contribute to the brown coloration in well-cured leaves. In the first place, much brown matter is extracted by cold water. Leaves thus exhausted will yield up another portion of brown matter to warm, dilute sulphuric acid, and finally still another portion[3] of a different chemical behavior is extracted by a warm, dilute solution of potassium hydrate.

---

[1] The expressed juice of a fresh tobacco leaf is at first without odor, but it gradually assumes that of fresh cucumbers, which later on is destroyed by putrefaction.

[2] According to Savery, the tannin of tobacco is identical with that of coffee. There exists, evidently, several kinds of phlobaphene, depending on the kind of tannin.

[3] This latter portion is a mixture of several compounds, some colorless and pectose-like, and one colored and phlobaphene-like. Twenty-five grams of fermented tobacco from Florida yielded 0.51 grams of this product. The cell membranes of the tobacco thus treated exhibit under the microscope a swollen appearance.

One author has assumed that the chlorophyll is first attacked in the curing process and destroyed, but this is not correct. The green color of the chlorophyll is in the beginning merely covered by the brown substances. In the thin samples of fermented leaves of a light-brown color green spots may frequently be noticed, and even dark-colored, freshly fermented leaves may sometimes yield a greenish solution upon extraction with strong alcohol. It is of some interest to note that the brown matters are insoluble in absolute alcohol.

*Ammonia.*—An interesting feature in the curing and fermenting process is the formation of a small amount of ammonia. As the green leaves contain some asparagin, the formation of ammonia might be due to a small extent to the decomposition of this amide, which readily yields ammonia and aspartic acid. But in certain tobacco crops there occur only minute quantities of asparagin. Certain amido compounds formed by decomposition of proteids and also a part of the nicotine in decomposing probably yield the principal amount of ammonia. The nicotine undergoes, in the fermentation process at least, a considerable diminution, as explained below.

The opinion that the ammonia deteriorates the quality of the product is certainly unfounded, as Fesca has correctly pointed out. It has been demonstrated by Behrens that during the curing process a part of the sulphur of the decomposed proteids is oxidized to sulphuric acid and that the amount of compounds soluble in ether decreases. The latter consist of a fatty substance and a volatile oil of disagreeable odor derived principally from the gland hairs.

The total loss of dry matter in the curing process is subject to great variation, depending mainly upon the amount of starch present at the time of gathering, as above stated. The diminution of dry matter may be as much as 40 per cent.

The principal changes in the curing process may be summed up as follows:

(1) Disappearance of starch.
(2) Formation of sugar and its partial disappearance by respiration.
(3) Decomposition of protein with formation of amido compounds.
(4) Decrease of fatty matter.
(5) Decrease of tannin.
(6) Change of color and flavor.

### THE SWEATING OR FERMENTATION PROCESS.

The so-called fermentation process develops in the tobacco leaves the characteristic qualities of the commercial article. It is natural to suppose that the same agency which finishes the curing process after the death of the cells remains active during the so-called fermentation process also. The fermentation follows immediately after the curing when both are done by the grower, but where the cured tobacco is bought up by manufacturers several months may pass before it is subjected to the sweating process. This operation begins when the tobacco is in the proper "order" or "case," being brought into this condition naturally on a damp day, or by an exceedingly cautious moistening, avoiding any visible water on the leaves. The amount of water applied must just suffice to bring on moderate imbibition. The total amount of water necessary to bring on a normal sweat is from 18 to 25 per cent of the moistened leaf. A portion of this water (about one-fourth) is again lost during the sweat.

The sweating of the Florida leaf in bulk requires from six to eight weeks, the original crude and rank smell of the cured tobacco being gradually changed to the proper aroma of the finished tobacco, and the glossy appearance and the texture[1] being well brought out. Light-colored wrappers require a slower and cooler fermentation than the dark-colored leaves used as dark wrappers or fillers.

*Rise of temperature.*—When the cured tobacco is sold by the farmer a large number of leaves are tied together at the base, forming "hands." At the beginning of the sweat such "hands" are well shaken in order to open all the foliage and admit air to every part. Then commences the moistening, when necessary, which is done by exposing the "hands," under continuous shaking, to a current of steam issuing from a pipe; by spraying with a fine spray; or by dipping, in which case the bases of the "hands" are plunged into water and shaken, the adhering water being soon drawn by capillary attraction into the leaf. These "hands" are then packed, with the butts outside, in piles 4 to 5 feet wide and 12 to 15 feet long. The rooms, which contain a large number of such piles, are kept warm, and steam passes freely from a number of pipes into the air of these rooms to secure uniform moisture, as otherwise the warming piles would soon become too dry. The temperature of these piles rises in from one to two days considerably above the temperature of the fermenting room and may reach 52° C. (126° F.) or higher. Repacking becomes necessary in from three to four days in order to check the rise in temperature and to shake out the leaves. The lower "hands" are now placed on the top and the outer ones in the center in order to give all leaves an equal chance to improve. The temperature now rises more and more slowly, the next repacking not being necessary before about seven or eight days. Altogether the piles are repacked from five to eight times. When the temperature rises too high the color or the aroma may be injured, hence frequent examinations are necessary. These examinations are made by pushing the hand into the piles, a decision being reached by the sense of feeling.[2]

*Oxidation.*—Tobacco manufacturers are well aware of the fact that a moderate quantity of air should gain access to the interior of the

---

[1] The texture, or grain, of the leaf means to tobacco manufacturers small points plainly visible on the extended leaf. It appears that these points are the bases of the gland hairs, most of which break off in the curing and sweating processes.

[2] For details relative to the treatment of the fermenting tobacco heaps the reader is referred to Farmers' Bulletin No. 60 and to the next report on tobacco.

fermenting piles and that undue pressure must be avoided in order not to diminish this access of air more than is necessary to insure an accumulation of heat. Not only are numerous little channels left naturally in the piles, but diffusion also will set in as soon as the air in the piles becomes warmer than the surrounding atmosphere.

Repeated efforts have been made to replace the sweating or fermentation process by a direct oxidation. Dr. Mew, of the Army Medical Museum of this city, assures the writer that some experiments made by him about twenty years ago to improve the cured tobacco leaf by direct application of a dilute solution of permanganate of potassium resulted in an essential improvement, the product being milder. Similar results have been recently mentioned by Kiessling.[1] In Germany a patent has been granted to the firm of Siemens & Halske for treating tobacco with ozone. However, oxidation often takes quite an undesirable turn, and the danger of destroying the aroma is quite as great, if not greater, than the likelihood of developing it by artificial means.

*Losses.*—Jenkins has shown[2] that the losses in fermentation are apparent in the nicotine, protein, amido compounds, nitrogen-free extract, and also, to a much less extent, in the ether extract. The loss of nicotine varies considerably in different samples and was found by Jenkins to range from one-sixth to one-half in three samples analyzed. Behrens observed in one sample a decrease of nicotine from 1.46 per cent in the cured leaf to 1.07 per cent in the fermented leaf. Dambergis found in air-dry Greek tobaccos, having from 7 to 14 per cent of water, from 2.8 to 0.7 per cent of nicotine.[3]

The question as to how much the loss of organic matter amounts to during the sweating process can be answered only approximately and by comparing parts of one and the same leaf, but a constant result will never be reached, as the nature of the proceeding in fermentation brings on differences in temperature, water content, and access of oxygen, and thus leads to variations. In the fermenting heap thick and thin leaves occur, often varying more than 20 per cent in weight for an equal surface area. Leaves grown in the shade are thinner than those exposed to direct sunlight, and in hot, dry summers the leaves are thicker and coarser than in moist, rainy seasons.[4] These conditions of course naturally influence the result.

Some tobacco manufacturers estimate the average loss during the fermentation process to be 15 per cent (organic matter and water together), while others estimate the loss of solid matter alone to be

---

[1] Der Tabak, Berlin, 1893.

[2] Conn. Agr. Expt. Sta. Ann. Rept., 1891.

[3] Oesterreich. Chem. Zeitg., No. 16, 1898.

[4] In the rainy season of 1891 Sumatra tobacco leaves weighed 52 grams per square meter, while in the dry season of 1892 the leaves grown on the same spot weighed 80 to 90 grams per square meter. Behrens explains this difference by the larger intramolecular spaces produced by excess of moisture (Landw. Vers. Stat., 1894, Band 43, p. 272).

from 4 to 5 per cent. According to Jenkins,[1] the losses may be even larger. He reports that "the upper leaves, short seconds, and first wrappers lost, respectively, by fermentation 9.7, 12.3, and 9.1 per cent of their total weight. But while three-fourths of the loss in the case of the short seconds consisted of water, in the case of the upper leaves almost three-fourths of the loss was dry matter. The first wrappers lost a little less dry matter than water."

*Development of gases.*—The formation of ammonia can be noticed by the characteristic odor in the fermenting rooms, but the amount is not so high as one might naturally be led to suppose from the intensity of the smell. About 3 liters of air from the interior of a fermenting pile when drawn through 25 cc. of Nessler's reagent, produced a light yellow color, indicating about 0.05 milligram of ammonia. No trace of hydrogen sulphide is given off. Test tubes containing filter paper moistened with basic lead acetate remained perfectly colorless for twenty-four hours in the fermenting heaps, hence it may be safely concluded that no protein decomposition resembling putrefaction takes place.[2] The amount of carbonic acid given off was also much smaller than would naturally be expected from the apparent energy of the action.

*Starch.*—Small quantities of starch are sometimes found in fermented tobacco when the curing process has not been carried out properly in all parts of the leaf or in parts of leaves broken, or injured by fungi, as observed by Müller-Thurgau, but this occurrence of a small percentage of starch interferes with the flavor just as little as does the closely related cellulose. The well-prepared tobacco wrappers from Florida examined by the writer did not show a trace of starch. The fact, however, that in the curing process the solution of the starch is going on with great energy forms a contrast to the observation that in certain cases remnants of starch remain unattacked during the fermentation process. This admits of hardly any doubt that the diastase[3] is gradually destroyed, perhaps by the proteolytic enzym.

*Sugar.*—As to the disappearance of the last remnant of sugar during the sweating, amounting, according to Müller-Thurgau, to from 1 to 3.3 per cent, some authors assume oxidation to carbon dioxide and water, and others assume a partial transformation to acetic acid. When it is taken into consideration that an alkaline medium can soon change glucose into organic acids (gluconic, saccharinic, etc.), especially in the presence of air, a more simple explanation would be at hand than

---

[1] Conn. Agr. Expt. Sta., Ann. Rept. 1892, p. 28. The leaves used for comparison were most carefully selected and were as nearly alike in color, size, and texture as possible.

[2] Nessler's comparison of the sweating process to putrefaction is certainly not admissible; neither is his declaration that the formation of ammonia is not normal, but simply a sign of true putrefaction.

[3] Diastase is absolutely necessary to dissolve and saccharify the starch. The dextrin and maltose thus formed may afterwards be transformed into glucose by the living protoplasm itself, wherever this latter comes under consideration.

the assumption of a perfect combustion of the glucose. There are organic acids present in the original tobacco leaf, such as citric, malic, and oxalic acids, in the form of neutral salts. A part of these acids may be changed and destroyed in the fermentation process, while other acids may be formed by the changes the glucose undergoes.[1] The nicotine is bound to organic acids and is not present in the free state; besides, most of the ammonia formed is in combination with organic acids, but a part of it is easily liberated by boiling the aqueous extract of the fermented tobacco. These vapors have a strong alkaline reaction and an ammoniacal odor, and are due either to the volatilization of some ammonium carbonate or to the dissociation of a neutral ammonium salt of a bibasic acid.

*Tannin.*—The amount of tannin, like that of nicotine, also decreases in fermentation. It varies from 0.3 to 2.3 per cent in commercial tobaccos. The Florida tobacco of 1898 contained only traces of tannin after the fermentation was over. The amount of fatty matter, or, more correctly speaking, of substances soluble in ether, was found by Jenkins[2] to decrease in fermentation from 3.5 to 2.8 per cent of the dry matter. Behrens observed in one case a decrease from 9.14 per cent in the cured to 8.34 per cent in the fermented leaf. The amount of such fatty substances was found to vary in different samples from 1.8 to 10 per cent and in some cases even more. The decrease of fatty matter during fermentation is probably due to the volatilization of a volatile ethereal oil. It is certainly very improbable that some true fat was oxidized to carbonic acid and water. Little attention has been given thus far to the small amount of resins in tobacco.

*Fiber.*—In regard to the fiber, Jenkins determined its amount in Connecticut tobacco as ranging from 13 to 14 per cent. Fesca and Jmai found the range in Japanese tobacco to be from 13 to 15 per cent. Only the ribs contained more—from 22 to 24 per cent.[3]

*Ashes.*—The amount of mineral matter is subject to very great variation, namely, from 10 to 27 per cent.

*Nitrate.*—A question of special interest is the fate of the nitric acid probably present exclusively as potassium nitrate in fresh leaves. Some authors believe that nitrification goes on during the fermentation process, which would lead to an increase of nitrate in the fermented leaf. This, however, has never been proved by chemical analysis and is indeed

---

[1] The precipitate obtained by copper acetate from a hot, aqueous extract of fermented tobacco contains, among other things, some succinic acid. The writer did not recognize butyric acid among the volatile acids in Florida tobacco, but acetic acid was present.

[2] Conn. Agr. Expt. Sta., Ann. Rept. 1890. Correct comparison is, however, possible only in calculating for a constant, e. g., cellulose.

[3] There are still certain substances in the fermented tobacco which thus far could not be characterized. Some analyses show from 1.7 to 18.9 per cent of nitrogenous extractive matter and from 8.6 to 16.7 per cent of indefinite insoluble matter. There exists great difficulty in isolating certain compounds from these mixtures.

highly improbable. Other investigators, for instance, Jenkins, have proved that the nitrate content undergoes only an insignificant decrease,[1] and still others, as Fesca and Behrens, assert that the nitric acid disappears completely, probably by reduction. These contradicting statements may be due to the great variations occurring in the nitrate percentage. A small amount might disappear completely while a larger amount decreases but little, although in both cases the absolute amount disappearing may be the same.

The writer examined, qualitatively, both fresh and fermented leaves from the same farm near Quincy, Fla., and found a moderate amount of nitrate in both. The samples of tobacco examined by Jenkins contained from 1.89 to 2.59 per cent of nitric acid ($N_2O_5$) at a water content of 23.5 to 27.5 per cent, while Behrens's samples contained only 0.2 per cent of this acid in the dry matter. From the disappearance of such a small amount of nitric acid, it can not be inferred that larger quantities would disappear entirely. In one of Jenkins's samples the amount of nitric acid diminished from 2.59 to 2.35, that is, a diminution corresponding to 0.4 per cent calculated for dry matter of the wrapper leaves. Dambergis observed variations of nitric acid in Greek tobacco of commerce of from 0.5 to 3.37 per cent of the dry matter. Not only the mode of manuring, but also the nature of the soil and the weather, influence the nitrate content of the plants, hence large differences can not be a matter of surprise.

The principal changes which take place during the sweating or fermentation process, as found by various investigators, may be summed up as follows:

(1) Decrease of nicotine.
(2) Increase of ammonia.
(3) Increase of alkaline reaction.
(4) Disappearance of sugar.
(5) Decrease of nitrate.
(6) Improvement of flavor and aroma.

### THE COLD SWEAT, AGING, OR AFTER-FERMENTATION.

The cold sweat which unfermented tobacco undergoes, and which corresponds with the aging of wines, may be intentionally carried on for as long as two years, where the main fermentation process has to be shortened for any reason or is not thoroughly completed. Fully and perfectly fermented leaves do not require this cold sweat, and the manufacturers of a good product prevent after-fermentation by giving such a degree of dryness in packing that further changes are stopped, as after-fermentation might finally lead to great differences in the product, which should be uniform. The interior of the piles or cases would naturally become warmer, and the leaves would change more

---

[1] Conn. Expt. Sta., Ann. Rept. 1892. To Jenkins belongs the credit of having first compared the fermented with the unfermented leaf in regard to the chemical changes.

than those near the sides. In many cases, however, a further change is found necessary, and slightly moistened sponges are placed in the cases with the tobacco in order to maintain a certain degree of moisture. After-fermentation carried on for too long a time might finally destroy all good qualities by further oxidations.

### THE PETUNING OF THE TOBACCO.

The petuning is an operation first practiced in Cuba, and consists in spraying a liquid on the leaves during or after the sweating process. The fillers only, and not the wrappers, are petuned, the intention being to give them a darker color, an improved flavor, and the appearance and character of a strong tobacco. The composition of the petuning liquid used in Cuba is kept secret, and indeed each planter claims to have something known only to himself. It is generally believed that one method of preparing the petuning fluid is by pouring organic fluids yielding ammonium carbonate over crushed tobacco stems, and letting this mixture digest. This liquid is, of course, very liable to putrefy, and consequently a most luxuriant growth of bacteria may be expected within a few days in the warm climate of Cuba. It is no wonder then that on the surface of Havana tobacco various bacteria are found, although it may be doubted whether they live long on these fermenting leaves. The ammonium carbonate contained in the petuning liquids increases the alkaline reaction already present in the fermenting leaf, and thus supports the energy of the oxidizing process which brings on the dark color frequently desired for the filler leaves.

One might naturally suppose that the ammonium carbonate would dissolve some resinous matter from the stems which would lead to an improvement of the aroma of the fillers if the hypothesis is correct that this aroma depends to a great extent upon the resin content of the tobacco plant, but this the writer holds is doubtful.

Petuning is practiced in some parts of the United States also, but the opinion of tobacco manufacturers whom the writer has consulted upon the subject is that the effect of the treatment is overrated. By the most intelligent growers a hot solution of ammonium carbonate is left to act upon the stems of Havana tobacco. This extract is prepared anew every day for use, which easily accounts for the fact that the tobacco leaves thus treated do not show any bacterial flora on their surface. The petuning liquid often has a different composition. The tobacco stems are extracted with water containing rum, molasses, or sour wine, consequently these liquids may swarm with bacteria after they stand for a while. The molasses is supposed to disguise the bitter taste derived from the stems.

Related to the petuning is the so-called "conditioning" of the tobacco, consisting in the spraying with a 2 per cent solution of glycerin. This operation is carried on only with chewing, plug, and cigarette tobaccos, and is intended to keep these products moist and pliable, as perfectly dry tobacco would easily crumble to a powder.

## THE BACTERIAL FERMENTATION THEORY OF SUCHSLAND.

It has long been recognized that the main feature of the processes going on in the sweating, or the so-called fermentation, of tobacco consists in oxidations. These are accompanied by certain decompositions liberating ammonia, and are the source of the striking development of heat in the fermenting piles. Now, what is the cause of these powerful oxidations? Nessler, as well as Schlösing, asserts that it is merely the common oxygen of the air that attacks certain compounds in the cells with great ease, no other cause being required. Schlösing admits bacterial action only for initiating the elevation of temperature, but not for the main processes later on. On the other hand, Suchsland attributes all the oxidations and the development of heat to the action of certain bacteria, which are specific for different kinds of tobacco and which impart to each of them a specific aroma.

Nessler's and Schlösing's views must assume substances of an unusual affinity for oxygen, if the rather indifferent atmospheric ogygen could exert such a powerful result without the intervention of any activifying principle, hence Suchsland's view seemed more probable and soon found many followers. He prepared pure cultures of microbes found upon different kinds of tobacco,[1] and by transferring those obtained from Havana tobacco to German tobacco he expected to develop the Havana aroma in the German tobacco, but thus far no new developments have startled tobacco growers. Davalos described mold fungi and microbes occurring upon fermenting tobacco leaves in Havana, but without proving their importance for the fermentation process (see Petuning). Vernhout observed only one kind of bacterium upon fermented tobacco leaves. This developed at 50° C. (122° F.) upon agar plates, and was a thermophile kind of the group of *Bacillus subtilis*.[2] It developed also in decoctions of tobacco and was capable of decomposing proteids with development of ammonia. Vernhout, however, leaves it entirely undecided as to whether this microbe plays any important part in the fermentation process. Also Koning[3] described several kinds of bacteria from fermenting tobacco leaves which he found to be identical with those occurring also on the green tobacco leaves. Besides the known bacteria, *B. mycoides* and *B. subtilis*, he described five aerobic new kinds, called *B. tobacci* Nos. I, II, III, IV, and V, of which *B. tobacci* III seemed to have most influence on the aroma. These statements may well be doubted, as a direct microscopical investigation of the surface of the fermenting leaves is wanting.

---

[1] Ber. d. Deut. Bot. Ges., Vol. IX, 1891.

[2] How some authors can assert that the fermentation is caused by *anaerobic* bacteria when it is a known fact that the most important changes going on consist of *oxidations* remains difficult to understand.

[3] Zeitschr. für Unters. der Nahrungs and Genussmittel, 1898, No. 3. It may also be mentioned that this author claims to have discovered the bacteria causing the mosaic disease of tobacco, while the most careful researches of Bejerinck have proved that bacteria are not the cause of it.

The writer has repeatedly tried to scrape off bacteria from the surface of freshly fermented Florida tobacco leaves, but has searched in vain with the highest magnifying power for the millions of microbes naturally to be expected if they really play a part in raising the temperature of the fermenting heap and bringing on powerful chemical changes. These *fermenting* leaves are, however, exceedingly *smooth* and *clean*, and the scrapings obtained from them consist almost exclusively of particles of the epidermis. Only here and there, by application of staining methods, some small globules become visible which might represent spores or cocci. Certainly so few microbes could never be held responsible for the action in the fermenting heap, but, on the contrary, colonies of luxuriant growth, as seen spreading profusely upon potatoes or agar, ought to be expected. It is very instructive that Behrens in his attempt to isolate bacteria from the surface of cured tobacco leaves, obtained two spore-forming microbes, *Bacillus subtilis* and a *Clostridium*. One can not suppress the supposition that both kinds have been present only as spores and as such would remain inactive during the so-called fermentation process, as there is not sufficient water to bring on their germination.

It is evident that for the proper examination of fermenting tobacco leaves one must *avoid petuned* leaves, upon which all kinds of microbes can be found when a putrefying petuning liquid is applied. But this liquid is not at all essential for starting the fermentation process. The fermenting Florida tobacco leaves the writer had under examination were not petuned, and he most emphatically declares (1) that there are no bacteria in the cells of the tobacco leaf, and (2) that the surface is remarkably clean and is not covered by a bacterial coating. This observation was made also by Mr. Albert F. Woods, of the Division of Vegetable Physiology and Pathology, two years ago in his study of spots on fermented leaves.

The chief object and pride of the tobacco manufacturer is to produce a cigar leaf of faultless quality. This would be impossible were bacteria to develop their activity promiscuously on the surface, as their first step would be to reach the nourishing material in the interior of the cells, otherwise they would be incapable of multiplying except for a short time. In gaining entrance to the cells the cellulose walls would have to yield, or, in other words, the surface of the leaves would be attacked.

We have here quite a different case from that of the fermentation of sauerkraut, which contains over 92 per cent of water and a proportion of cellulose to water as 1 to 62. In the fermenting tobacco leaf the amount of water is generally below 25 per cent and the proportion of cellulose to water is generally less than 1 to 1.5. In fact, the water present merely suffices to impregnate the cellulose walls and contents of the cells, and is entirely insufficient to bring organic matter from the interior of the cells to the surface, where bacteria might feed upon it.

It is indeed a matter of interest to observe how the tobacco leaf becomes less fit to support bacterial life after being cured and fermented. While the expressed juice of the fresh tobacco leaf exposed to the air at the ordinary temperature teems with myriads of bacteria within twenty-four hours, the equally concentrated extract of cured or fermented leaves will remain perfectly clear for many days.[1] On the fresh tobacco leaf, as everywhere in nature, numerous kinds of microbes occur, but these seem to die off when cured leaves are fermented, as will be seen from the following experiment by the writer with Florida tobacco leaves: Into about 15 cc. of sterilized beef broth, contained in three test tubes closed with cotton plugs, were introduced, with all necessary precautions, (1) a small scrap of fresh leaf, (2) a small scrap of cured leaf which had been packed two months waiting fermentation, and (3) a scrap of fermented leaf. The tubes were kept at from 15° to 18° C. for several days. No. 1 was turbid after one day, when a scum formed and the liquid became very turbid; the liquid swarmed with bacteria, thick and thin rods and cocci being revealed by the microscope. Nos. 2 and 3 remained perfectly clear, and after eight days merely a trace of flocculi was seen at the bottom, in which a few cocci (Sarcina (?)) could be recognized. Indeed, the juice of the fermented tobacco leaf acts as an antiseptic upon the ordinary bacteria of putrefaction. When a slice of meat is wrapped in a fresh tobacco leaf, and another in a moistened, fermented tobacco leaf, it will be seen after a few days that the former slice is rotten and the latter not.[2] This property of course disappears upon considerable dilution of the juice, as will be seen from the following experiments: Ten grams of fermented and well-dried tobacco leaf were pulverized and extracted with 250 cc. of boiling water. A part of the filtrate received an addition of sugar and another part an addition of peptone. The well-sterilized flasks were infected with small chips of fermented tobacco leaf and some of them kept at 50° C. (122° F.) for three days, and some at from 18 to 20° C. (64.4° to 68° F.). There was more or less development in all the flasks, but further tests, as the inoculation in peptone solution or on potatoes, revealed as the only organism a bacillus resembling *B. subtilis*. The development of the colonies, the mode of growth on the surface of peptone solution and on potato, and the spore-forming threads left no doubt on this point. This result is then in full accordance with the observations of others—that is, that this bacillus can be cultivated from fermented tobacco leaves. But as the most careful searching of the surface of the fermented leaves for the bacillus itself proved vain, it must be assumed that it exists on these leaves only in the form of spores.

---

[1] It may be mentioned, however, that a diluted (1 per cent) solution of a neutral nicotine salt will permit a bacterial growth.

[2] Southern manufacturers assert that the workmen in tobacco factories better resist epidemics than those not so employed.

When by accident leaves are too much moistened before they are subjected to the sweating, they will soon lose their coherence and show spots and finally holes. The water content, the writer found, in one such case amounted to 36 per cent. Here, then, is a true action of bacteria, which can develop under these conditions, as the high percentage of water admits an abundant exit of organic compounds from the interior of the cells to the surface and the formation of a diluted solution. Now, it is the experience of every tobacco manufacturer that the product will invariably spoil when the water content is increased to such a point as to permit an exit of soluble organic compounds from the cells. Here, then, begins the parallelism to the fermentation of sauerkraut or ensilage, but not before. The objection that certain kinds of thermophylic bacteria might be capable of developing on the leaves in the presence of a smaller percentage of water can not be sustained, as they require liquid food as well.[1] And how will they reach the interior of the cells without eating through the cellular walls, that is, without ruining the product? The claim that it is not the bacteria but the enzyms they produce that enter can not hold good, as the latter must be dissolved before they can migrate into the interior of the cells, and hence a water increase is again required. The conclusion that must invariably be reached, therefore, is that the bacteria found upon the fermenting tobacco leaves do not participate in any way in the fermentation process, but that they are accidentally present and probably only in the form of spores.

### THE OXIDIZING AGENCY IN THE FERMENTING TOBACCO LEAF.

After showing that the bacterial theory of Suchsland is erroneous, as there exists no bacterial coating on the leaves, the question naturally presents itself, what is the cause of the oxidizing action? The assumption of Nessler and of Schlösing that the contact with the atmospheric oxygen would suffice can not be correct for the following reasons: (1) The substances undergoing oxidation (tannin, nicotine, etc.) do not show such powerful affinities for oxygen as to account for the considerable development of heat; and (2) neither curing nor fermentation sets in when the fresh leaves are killed by direct application of steam, although those organic matters which become oxidized in the fermentation process are not changed at all thereby.

Neither the tannin nor the nicotine of the leaves can be energetically oxidized by the molecular oxygen of the air without assistance or stimulation of some sort. In the same way dilute alcohol can not be oxidized into acetic acid by the common molecular oxygen of the air except through the intervention of certain bacteria or platinum black.[2]

---

[1] Cohn made investigations on the growth of thermogenic micrococci in the refuse from the cotton purifier. However, he had to add a fair percentage of water to start the development.

[2] Müller-Thurgau declares (l. c., p. 508) that "In the beginning of the curing the changes consist in an increased respiration, but later on, after the cells have died, in other ('*anderweitigen*') oxidation processes," but he gives no explanation of the cause of these latter ones.

The oxidations in the treatment of tobacco commence with the curing process and are continued in the fermenting process. In the latter case, but not in the former, the aid of bacteria has been invoked for explanation. But when oxidations can go on in the *curing* without bacterial aid, even after the death of the cells, then it might be supposed that the same cause would also lead to oxidation later on during the fermenting process. Now, what is the true cause of these phenomena? There remains, in fact, as the only explanation the writer's suggestion that an *oxidizing enzym* is the final cause of the energetic oxidizing action after death of the cells as it is capable of instigating certain compounds to take up the molecular oxygen of the air.

The formation of enzyms is a physiological necessity for every living organism. Various enzyms come into action especially in the development of shoots, as well as in the inanition state of the plants. Green plants, as well as lower fungi, prepare enzyms, which may act on protein, polyanhydrids of glucoses, glucosides, or fat, splitting or dissolving these bodies and thus making them more easily accessible to the protoplasm.[1] The list of enzyms has been enlarged in recent years by the oxidizing enzyms or the oxidases, which were brought to our knowledge first by French savants, as Gabriel Bertrand, Bourquelot, Gouirand, Cazeneuve, and others. The most thorough investigations on this subject are those by Professor Bertrand, who has shown their wide distribution through the vegetable kingdom.[2]

The best reaction for oxidizing enzyms consists in the production of a blue color with the tincture of guaiac, which reaction can be obtained with various vegetable objects. This blue coloration is produced in many cases only upon addition of peroxide of hydrogen, in which case it was formerly considered as a reaction upon diastase. While the crude diastase of malt gives this blue reaction in a very marked degree, the diastase of certain fungi (*Aspergillus oryzæ*) will, as the writer long since ascertained, yield this reaction either only slightly or not at all, although this diastase has very energetic qualities and produces glucose from starch. Raciborski and other authors also have proved that this blue reaction is not characteristic of pure diastase, but only of an admixture of an oxidase.

The brown, black, or reddish coloration of freshly prepared juices of potatoes, turnips, etc., setting in when exposed to the air, the brown color of the falling leaves in autumn, and similar phenomena are generally due to the action of the oxidases. The oxidation of tannin by oxidases plays an important part in certain fruits ripening or overripe.

---

[1] A trophic irritation is exerted when rapidly developing cells or cells in a state of inanition require nourishment, and this stimulus leads to the production of enzyms by the nuclei—an interesting case of physiological adaptation. Well-nourished cells killed in the full vigor of life often give only slight indications of amylolytic and proteolytic enzyms.

[2] Further contributions have been published by Grüss and by Raciborski (Ber. d. Deut. Bot. Ges., Vol. XVI, Nos. 3 and 5, 1898).

The blackening of bananas some time after they are gathered and the brown color on the surface of a slice of apple may also be mentioned as due to these agents.

Oxidizing enzyms also occur in animal organisms, as the investigations of Pieri, Abelous, Bougault, Salkowski, Yamagiva, Linossier, Jaquet, and Schmiedeberg have revealed. Such enzyms were found in various organs, and are capable of easily oxidizing not only guaiac tincture,[1] but also certain aldehydes, such as salicylic aldehyde. Spitzer has determined the amount of oxygen liberated by different organs from peroxide of hydrogen, and has observed that various poisons, such as potassium cyanide, hydroxylamin, etc., small quantities of acids and alkalies, and a temperature of about $70°$ C. ($158°$ F.) destroy or diminish the oxidizing action of this animal enzym. Water extracts the enzym from the organs, and highly diluted acids, cautiously added, precipitate it with all its original properties. (Abelous could, however, prepare clear solutions only by application of potassium nitrate.) It has the character of a nucleo-proteid and contains from 0.19 to 0.23 per cent of iron. On the other hand, Bertrand and Villiers have found a small amount of manganese in the vegetable oxidases.

That oxidations also can proceed in certain cases without the aid of oxidizing enzyms is a well-known fact. But this is only the case with substances of a specific kind showing a great chemical energy, and even in such cases the presence of oxidizing enzyms will cause such a powerful increase of intensity that the difference becomes most striking, especially when chromogens consisting of certain derivatives of polyvalent phenols are present. The colored product (brown, red, or black) formed by oxidation will appear much sooner and in much greater quantity in the presence of certain oxidizing enzyms than in their absence.[2] On the other hand, oxidizing enzyms can bring on oxidations with certain compounds, as, for example, tyrosin, which under ordinary circumstances would not be oxidized at all by the indifferent oxygen of the air.

Bertrand has characterized different oxidases. While the oxidase of *Rhus vernicifera*, the Japanese lac tree, oxidizes mainly benzene derivatives containing at least two hydroxyl or two amido groups in ortho or para position, another oxidase, isolated from certain green plants, as well as from fungi, acts easily in tyrosin, which the former can not affect, therefore he distinguishes the latter as tyrosinase from the former as laccase. It is the laccase which acts upon the laccol in the

---

[1] This blue reaction can be obtained not only by the action of oxidizing enzyms, but also by that of powerful oxidizing chemicals. Such bodies (nitrous acid, free chlorine, etc.) are usually absent when the test for oxidizing enzyms is made. Any intelligent chemist will be able to decide at once by control experiments whether he can trace the reaction rapidly setting in to oxidases or not.

[2] The attempt to explain such rapid oxidations by the assumption that certain salts or ordinary albuminous matter would activify the oxygen, must be considered a failure.

juice of the lac tree and converts it by oxidation into a black substance. Laccase is killed at 75.5° C. (168° F.) and gives the guaiac reaction without the aid of hydrogen peroxide. Like laccase, the oxidases of *Senecio vulgaris*, *Lactuca sativa*, and *Taraxacum dens leonis* fail to attack tyrosin. In certain objects, especially in fungi, however, laccase and tyrosinase occur simultaneously.

The oxidation of polyvalent phenols by laccase leads not only to organic acids, but even to the production of carbonic acid. Bertrand observed in one case that for 23.3 cc. absorbed oxygen as much as 13.7 cc. of carbonic acid were produced.

Gouirand has observed in certain spoiled wines an oxidase[1] which oxidizes the coloring matter, the tannin, and the alcohol of the wine, with production of carbonic acid. This oxidase is destroyed in plain aqueous solution at 72.5° C. (162.5° F.), while in the wine a temperature of 60° C. (140° F.) suffices. Very small doses of sulphurous acid will also kill it. It is supposed to be derived from the fungus *Botrytis cinerea*, which frequently grows upon ripening grapes,[2] while Martinand and Tolomei observed an oxidase in ripe grapes. As to the Florida tobacco leaf, the writer has demonstrated the presence of a relatively large amount of oxidases in it.

These oxidizing enzyms belong, like other enzyms, to the protein compounds, forming a special group of labile proteins, i. e., proteins containing much chemical energy, which on the one hand is the cause of their activity and on the other of their changeability to indifferent proteids by heat, acids, and poisons. They are, as it is expressed, easily killed. The labile, active atomic groups in the molecules change thereby, the atoms migrating into more stable position.

*Views on the physiological functions of the oxidizing enzyms.*—As to the physiological function of the oxidizing enzyms, no perfectly satisfactory explanation has thus far been proposed. Some authors suppose that they are important agencies in the respiration process and that even respiration itself is caused by them when they are supported by certain properties of the living protoplasm. This is, however, improbable for several reasons: (1) Not every plant contains oxidizing enzyms; (2) many plants contain them only in certain stages; and (3) carbohydrates and fat, the materials which by their combustion serve for support of the respiration and for the production of energy, are not attacked by the oxidizing enzyms, but are attacked very energetically by the protoplasm.

Portier believes that the oxidase of the blood, of which he made a special study, serves to augment the vitality of the leucocytes, which prepare the oxidase and finally deliver it up to the blood when they die. This hypothesis will certainly not find support. The suggestion has also been advanced that the oxidizing enzyms play in plants the

---

[1] This œnoxidase is supposed by Bertrand to be identical with laccase.
[2] See also Cazeneuve, Compt. rend., Vol. CXXIV.

same part that the hæmoglobin does in animals, but neither is this view justified, as the oxidizing enzyms are not carriers of molecular oxygen, but simply instigators of oxidation.

The writer's view on this subject is that as the living protoplasm can oxidize carbohydrates and fat, but does not attack or attacks only with difficulty compounds of the benzene group, and, on the other hand, as just the opposite takes place with the oxidizing enzyms, it may be inferred that there exists between the protoplasm and the oxidizing enzyms a certain division of labor, the former oxidizing the compounds of the methan series and the latter those of the benzene series. The former provides for the kinetic energy of the cells; the latter destroys by partial oxidation noxious by-products. The oxidations in the former case are generally complete, but in the latter only partial.

The oxidizing action of enzyms might be compared to that of platinum black. In both cases chemical energy is conveyed to certain organic compounds, which are thus rendered capable of taking up the oxygen directly from the air. The further inference might also be justified that just as platinum black brings on not only oxidations, but also reductions under certain circumstances, the same may be possible for the oxidases; for example, if platinum black is added to a mixture of glucose and potassium nitrate in aqueous solution, a reduction of nitrate to ammonia takes place by aid of hydrogen atoms in the sugar, while the oxygen of the nitrate is thrown upon the glucose and organic acids thereby formed. When the analogy of action of the oxidase to platinum black is justified, there will be a simple explanation for the disappearance of a certain portion of the nitrate and also of a certain portion of the glucose during the fermentation process of tobacco. Preliminary qualitative experiments by the writer have indeed proved the formation of ammonia under these conditions. A full account of quantitative tests will follow in a later bulletin.

The oxidizing enzyms may occur in various parts of the plant—in young and active as well as in dormant tissue. Grüss has observed that there occurs frequently, but not always, a coincidence between the transformation of starch and increase of oxidase. Whether the amount of oxidase augments with the ripening of fruits has not been thoroughly investigated. Tolomei observed that in olives it does increase during the ripening process.

The juice of the fresh tobacco leaf soon turns dark upon exposure to air and gradually forms a sediment, but if boiled this dark coloration does not set in, the oxidase having been killed.

*The tobacco oxidase and peroxidase.*—There exist, evidently, two kinds of oxidizing enzyms in the Florida tobacco leaf. The first kind oxidizes guaiaconic acid (the characteristic reactive in the guaiac resin) to guaiac blue without the aid of peroxide of hydrogen, but the second kind oxidizes it only when this substance is present. Both kinds of oxidizing enzyms, which may be distinguished as tobacco oxidase and tobacco peroxidase, occur in the fresh as well as in the recently fermented Florida tobacco leaf.[1] The former enzym is, however, much more sensitive to heat than the latter, being killed at from 65° to 66° C. (149° to 151° F.), while the latter is killed only at from 87° to 88° C. (188.6° to 190.4° F.).

Dried tobacco leaf (not cured) was finely pulverized, and 1 gram of the powder left with 20 grams of water for one hour at the ordinary temperature. A part of the filtrate was heated for three minutes to 55° C. (131° F.). To about 2 cc. of this liquid were then added a few drops of tincture of guaiac,[2] whereupon a blue coloration appeared in a few minutes, exactly as in the control case. A second portion was now heated to 65° C. (149° F.) for three minutes, and the test applied after cooling, but only a slight trace of a blue color was noticed after ten minutes. Evidently most of the tobacco oxidase was destroyed at that temperature.

The killing temperature of the tobacco peroxidase was determined in a similar manner. However, here reaction is still obtained with great intensity after the solution is heated for three minutes to 80° C. (176° F.), but very feebly after heating for one minute to 87° C. (188° F.).

Another reaction for oxidizing enzyms is the so-called indophenol reaction, consisting in the production of a blue color when an alkaline solution of $\alpha$ naphthol with paraphenylendiamine is acted upon by an oxidase. This reaction must, however, quickly set in and with great intensity, otherwise no reliable conclusion can be drawn. Cured and fermented tobacco from Florida did not show this reaction in a marked manner,[3] but it set in at once upon the addition of a little peroxide of hydrogen. The latter alone will not produce this result in the absence of an oxidizing enzym.

In the manner above mentioned the writer's investigations have shown that dark tobacco two years old, from Quincy, Fla., yielded no reaction for tobacco oxidase, but still a moderate one for tobacco peroxidase, while a sample of light-colored tobacco four years old from the same source yielded not the slightest reaction either for the oxidase or the peroxidase. Evidently these enzyms themselves are gradually changed. From these observations it may be inferred that the cold sweat, or after-fermentation, might thus proceed for about two years and end by the gradual dying off of the oxidase and peroxidase.

---

[1] French savants were the first to call attention to this difference between the oxidizing enzymes. The names oxidase and peroxidase, proposed by a French savant, are not specific names, but group names. There may exist among various oxidases and peroxidases as many differences as there are among protein bodies. Hence it is entirely unjustifiable, at this stage of our knowledge, to introduce one specific name for all peroxidases, as one author has done.

[2] In all these cases freshly prepared guaiac tincture (1:50) was employed, as old guaiac tincture is unreliable and with peroxide of hydrogen alone will sometimes yield a greenish coloration.

[3] Only a slow and weak reaction was thus developed.

When the fresh leaf of the tobacco is rapidly dried at about 60° C. (140° F.) and then moistened again and kept in a moist atmosphere, the veins and their finest ramifications turn brown in about half an hour, while the mesophyll and epidermal cells remain green even after a week. Further investigations on this point will be made later. In the fresh leaf, both oxidizing enzyms, the oxidase and the peroxidase occur in the ribs and veins as well as in the parenchyma, the indications being that they are more abundant in the ribs than in the parenchyma. The bundle sheath and sieve tissue give the most intense reaction on the oxidase, while the reaction on the peroxidase sets in quickly and with about uniform intensity in all the cellular tissues. The growing point and youngest leaves contain an especially large quantity of the oxidase. A section through the stalk shows oxidase only in the sieve tissue and bast parenchyma, while peroxidase also is contained in the pith.[1] Both enzyms are found in the root, the former more in the central and the latter in the peripheral parts and also in the flower. The stigma of the pistil and the stigmatic fluid also show strong reaction upon oxidase.

The two oxidizing enzyms are also contained in the young tobacco plants. Several dozen of these, measuring on an average not more than 3 to 4 cm. from the tip of the root to the plumula, were rubbed in a mortar with a little water and some sand. The filtrate gave a very intense reaction for oxidase,[2] and after this was destroyed by warming to 70° C. an intense reaction for peroxidase also.

A colorless clear solution of the tobacco peroxidase can be obtained in the following manner: A number of fresh tobacco leaves are well crushed in a mortar, with the addition of sand and some dilute alcohol of 30 per cent. This mixture is pressed and the turbid liquid directly mixed with three times its bulk of strong alcohol. After standing two hours the mixture is thrown upon a filter and the filter contents, after being washed with some alcohol, extracted with about four to six times its bulk of water at the ordinary temperature, heated for a minute to 70° C. (158° F.), and filtered. This clear, colorless filtrate gives no indication of the oxidase, but a very intense reaction for peroxidase. When this solution is compared with that of the juice of fresh tobacco leaves it is easy to decide what result is caused by the oxidase alone.

To determine whether the tobacco oxidase bears more resemblance to the tyrosinase or the laccase a few drops of freshly expressed juice of normal tobacco leaf were added to 2 cc. of a cold saturated tyrosin solution, but even after four hours no characteristic darkening of the

---

[1] In order to observe the localization of the peroxidase small pieces of the tissue are treated with strong, but not absolute, alcohol for three minutes at 70° C. (158° F.). Thus the oxidase is killed and can not interfere with the tests for the peroxidase.

[2] The indophenol reaction did not turn out satisfactorily, only a weak violet-blue color resulting. On the addition of hydrogen peroxide, however, an intense blue reaction was at once obtained.

mixture was observed. In this regard, therefore, that oxidase would resemble laccase more than it would tyrosinase.[1]

To extract oxidases from fermented or cured tobacco as completely as possible it is necessary to thoroughly pulverize the samples and to let the water act for some time at from 20° to 30° C. (68° to 86° F.) before filtering. After complete drying, the samples can be easily pulverized very fine. The following experiment proves that when the tissue is not pulverized the peroxidase is but very imperfectly extracted, the passage through the cellular walls being quite slow. Fermented tobacco leaves were three times soaked in water and the brown liquid pressed out, the first soaking lasting half an hour and the second and third soakings five minutes each. Although the sample was thus nearly exhausted, it nevertheless yielded, when left with some alcohol of 30 per cent for one day, a light-colored liquid with a very intense reaction for peroxidase.

It may safely be assumed that in the majority of instances the oxidase will prove the more energetic of the two oxidizing enzyms. For example, its action upon pyrocatequol and hydroquinone is much more energetic than that of the peroxidase. On the other hand, however, the former succumbs much more quickly to noxious influences, e. g., the action of alcohol or rising temperature.

The fact that the peroxidase forms guaiac blue from guaiaconic acid with the aid of hydrogen peroxide only does not indicate that its oxidizing action in every case depends upon the presence of the latter.[2] The peroxidase can, on the contrary, also exert oxidizing action upon various compounds without the assistance of hydrogen peroxide. Thus, para-amidophenol is gradually changed by it to a dark brown substance.

*Hydroquinone* in dilute solution gradually assumes a reddish color in the presence of the peroxidase, but in its absence there is scarcely a trace of coloration within twenty-four hours.

*Pyrocatechol* is scarcely attacked by the peroxidase within twenty-four hours, but on a further addition of a little hydrogen peroxide it turns to a dark brown in five minutes. Hydrogen peroxide added alone does not act thus.

*Pyrogallol* is slowly attacked by the peroxidase and turns brown in twenty-four hours. The oxidase acts also in this case much more energetically than the peroxidase.

*Tannin* solution shows in twenty-four hours a yellow color in the presence of the peroxidase, but in the control solution merely a slight

---

[1] The peroxidase of tobacco on the other hand bears resemblance to the peroxidase of pus described by Linossier.

[2] The hydrogen peroxide is decomposed by enzyms into water and oxygen, but this oxygen in *status nascens* is charged with more chemical energy than the free oxygen of the air, i. e., the two atoms constituting the molecule are for a time in a more energetic motion than in the latter case, hence the action of the oxidizing enzyms is facilitated by this nascent oxygen.

trace of coloration is perceptible. The addition of a little hydrogen peroxide to both will increase the difference of coloration still more.

All these tests were made at from 18° to 20° C. (64.4° to 68° F.). A number of other compounds were also tested, such as arbutin, guaiacol, and toluidine, but no decisive reaction was obtained within twenty-four hours at the ordinary temperature.

There may exist great differences in the amount of tobacco oxidase and tobacco peroxidase produced in different varieties of the tobacco plant and under different conditions. The quantity of each may even differ in the upper leaves fully exposed to the sun and the lower leaves growing mostly in the shade. There may also be formed compounds in certain varieties of tobacco that will more quickly destroy the enzyms during curing, or fermentation, than in other varieties. Thus considerable difference was noticed in comparing a sample of tobacco from Connecticut with one from Florida. In the fermentation of the former the tobacco peroxidase was almost completely destroyed, while in that of the latter a considerable part was still intact. Moreover, neither the fermented nor the cured Connecticut leaf contained any tobacco oxidase, although it was found in a greenhouse specimen of the fresh leaf.

It is interesting to note that the best way of bringing the oxidizing enzym to the fullest action possible is that practiced in the curing of the Perique tobacco.[1] The rolls, or twists, of the tobacco leaves are subjected to a pressure of about 7,000 pounds per square foot to bring the juice from the interior of the cells to the surface. After twenty-four hours the tobacco is taken out and aired a few minutes, which causes a darkening to set in. In this way the juice is reabsorbed by the tissues, whereupon the pressure is again applied. This operation is repeated daily for ten consecutive days, and at longer intervals thereafter. A very dark product is thus obtained, but it is not strong, as the oxidation of the nicotine has been carried very far.

One of the most interesting features of the sweating of tobacco is the destruction of a part of the nicotine, this part yielding up its nitrogen probably as ammonia, which is indeed a product of the sweating. It was, of course, of importance to prove that the oxidizing enzyms contained in the tobacco leaf can decompose nicotine, and for this purpose 50 grams of cured tobacco from Connecticut which had not yet been subjected to fermentation and showed a strong reaction for peroxidase, but none for oxidase, was thoroughly moistened with water. After two hours 250 cc. of alcohol of 50 per cent was added and the mixture allowed to stand for two days. The liquid obtained by pressing was now mixed with one and a half times its volume of absolute alcohol and the brown-colored precipitate washed upon the filter with some alcohol. After pressing between filter paper, the precipitate, containing a large proportion of the peroxidase, was dissolved

---

[1] Farmers' Bull. No. 60, U. S. Dept. of Agr.

in 20 cc. of water and 0.5 gram of nicotine tartrate added. This mixture was digested for two days at from 50° to 60° C. (122° to 140° F.) in a large flask, holding about 500 cc. of air, to enable oxidation to go on. An addition of a small amount of thymol prevented bacterial growth. A small U tube, holding 10 cc. of dilute chemically pure sulphuric acid of 0.2 per cent, was attached to the flask. The examination of this acid after two days with Nessler's reagent indicated that ammonia was present, but the colorimetric comparison showed that the amount was hardly more than 0.1 milligram. However, the amount of ammonia formed but not volatilized was much larger, as indicated by the strong reaction obtained after the addition of a little potassium carbonate to the mixture and warming for a short time in order to liberate the ammonia in the form of carbonate from other less volatile salts.[1] Thus there can be no doubt that the tobacco peroxidase can attack nicotine with formation of ammonia, but this process is exceedingly slow. Indeed, the sweating, lasting fully eight weeks, can diminish the nicotine content on an average only by about one-third.[2]

What the products of destruction of nicotine are besides ammonia can be determined only when the purified enzyms and a pure nicotine salt serve in large quantities for the experiment. It may be mentioned, however, that the writer has examined in vain an aqueous extract of fermented tobacco for nicotyrin and nicotinic acid—known oxidation products of nicotine.

The writer has now fully established the presence of oxidizing enzyms in the tobacco leaf.[3] That such enzyms can exert a powerful action upon certain compounds, leading even to the formation of carbonic acid, is known.[4] Oxidations produce heat, hence it can safely be inferred that the so-called tobacco *fermentation* consists in the activity of oxidases, while the *curing* of tobacco consists in the combined work of oxidases, diastase, and peptase. As the use of the term "fermentation" might lead in this case to an entirely erroneous conception, the writer proposes "oxidizing enzymosis" or "oxidizing enzymation" as correct scientific designations.

There has already been mentioned an interesting case of oxidase action in a technical branch, viz, the preparation of the Japanese lac. Furthermore, in the manufacture of the natural indigo bacteria are not concerned (Molisch), but simply an oxidizing enzym (Bréaudat).

---

[1] The oxidase might have exerted a more powerful action on the nicotine than the peroxidase.

[2] A control experiment was made with a colorless peroxidase solution (p. 29) upon highly diluted free nicotine at the ordinary temperature, in order to observe whether a brown solution is produced by a change of the nicotine, but the mixture remained colorless after one day under these conditions. However, it may be mentioned that nicotine, when exposed a long time to air and light, will turn brown.

[3] He has already pointed out (p. 15) that there is sufficient access of air possible to enable oxidation in the tobacco piles.

[4] The further inference is certainly justified that certain basic compounds might thus give up their nitrogen in the form of ammonia.

Another instance where an oxidizing enzym plays a part in a technical branch is the "fermentation" of the olive, which is practiced in certain parts of Italy. It is believed that by this operation an oil of superior quality is obtained and that the yield by pressure is larger, but this has not been confirmed. It has been shown by Tolomei[1] that this olive "fermentation" is due to the action of an oxidizing enzym, to which also is due the fact that olive oil is bleachable by sunlight.

When the freshly gathered olives are kept in sacks their temperature gradually rises far above that of the rooms, oxygen is absorbed, and carbonic, acetic, and sebacic acids, and small quantities of the higher volatile fatty acids are formed. This process goes on to a larger extent when a temperature of 35° C. (95° F.) is reached. These changes do not occur if the olives are kept in nitrogen or carbonic acid gas; neither do they occur when the olives have been heated to 75° C. (167° F.) for forty-five minutes. For obvious reasons the spontaneous rising of temperature is noticed only when a large number of olives are kept together. Tolomei showed that the oxidase extracted with water and purified by a repeated precipitation with alcohol produces guaiac blue from guaiac tincture, forms purpurogallol from pyrogallol without the aid of peroxide of hydrogen, quinhydrone from hydroquinone, and a brown substance from gallic acid. He calls this oxidizing enzym olease. As unripe olives do not contain this olease, the oil pressed from them will not bleach upon exposure to the sunlight, but will do so after being shaken with an aqueous extract of the ripe olives. On the other hand, olive oil will sooner acquire rancidity in the presence of the olease than when free from it.

Finally, still another case may be pointed out where oxidases might possibly play a part—that is, in the so-called fermentation of the cacao beans, by which a bitter principle is destroyed.

### SUMMARY.

(1) The so-called tobacco fermentation is not caused by bacteria.

(2) The amount of water present in normally fermenting tobacco leaves is insufficient to bring nourishment for the microbes from the interior of the cells to the surface of the leaves. It barely suffices for imbibition of the organic matter.

(3) There exists no bacterial coating on the fermenting tobacco leaves under normal conditions, but some spores may occur.

(4) In the so-called petuning of tobacco an immense number of bacteria may be transferred to the leaves. These bacteria, however, are not essential for the fermentation, but on the contrary, may prove noxious as soon as a small surplus of water enables them to further develop.

(5) Suchsland's theory that the aroma of tobacco is caused by specific bacteria is incorrect.

(6) The principal changes that take place during the curing and fermenting of tobacco are due to the action of soluble ferments or enzyms.

(7) Several kinds of enzyms act in the curing process, (a) an amylolytic, (b) a proteolytic, and (c) two oxidizing enzyms, while in the fermenting process the main changes are due to oxidizing enzyms alone, and consist in the oxidation of nicotine and other compounds.

(8) The presence of the amylolytic and the proteolytic enzyms is inferred from the saccharification of the starch and the decomposition of proteids, but the enzyms themselves have not yet been isolated from the tobacco leaf.

(9) In green tobacco two oxidizing enzyms may exist, an oxidase and a peroxidase. The former succumbs much more readily to noxious influences than the latter and in all probability exerts a more powerful action.

(10) The development of color and aroma is due principally to the action of the oxidizing enzyms.

### RECENT FOREIGN LITERATURE.

*Nessler*, Der Tabak, etc., Mannheim, 1867.

*Schlösing*, Sur la Fermentation en Masses du Tabac, 1888 and 1889.

*Müller-Thurgau*, Ueber das Verhalten von Stärke und Zucker in reifenden Tabaks-Blättern, Landw. Jahrb., 1885.

*Fesca and Jmai*, Ueber Cultur, Behandlung, und Zusammensetzung japanischer Tabake, Landw. Jahrb., 1888.

*Wagner*, Tabak-Kultur, etc., Weimar, 1888.

*Behrens*, Weitere Beiträge zur Kentniss der Tabakspflanze, Landw. Vers. Stat., 1894 and 1895.

*Kiessling*, R., Der Tabak, etc., Berlin, 1893.

*Mayer*, Adolf, Der Tabak, Landw. Vers. Stat., 1891.

*Suchsland*, Ueber Tabak-Fermentation, Ber. d. Deut. Bot. Ges., Heft IX, 1891.

*Davalos*, Nota sobre la Fermentacion del Tabaco; Abstr. in Centralbl. f. Bakt., Heft XIII, 1893.

*Vernhout*, Rapport over het Bacteriologisch Underzoek ven Gefermenteerde Tabak, 1898; Abstr. in Centralbl. f. Bakt., Heft II, Abt. IV, p. 778.

*Behrens*, Die Beziehungen der Mikroorganismen zum Tabaksbau und zur Tabak-Fabrikation, Sammel-Referat, Centralbl. f. Bakt., 1896, Heft II, Abth. II, p. 514.

www.ingramcontent.com/pod-product-compliance
Lightning Source LLC
Chambersburg PA
CBHW062234220526
45471CB00009B/3480